MERCURY

by Emma Bassier

Cody Koala

An Imprint of Pop!
popbooksonline.com

abdobooks.com
Published by Pop!, a division of ABDO, PO Box 398166, Minneapolis, Minnesota 55439.

Printed in the United States of America, North Mankato, Minnesota.

102020
012021

THIS BOOK CONTAINS
RECYCLED MATERIALS

Cover Photos: NASA, Mercury; iStockphoto, background
Interior Photos: NASA, 1 (Mercury), 9 (bottom left), 9 (bottom right), 14, 17; iStockphoto, 1 (background), 5, 6, 9 (top), 19 (top); Shutterstock Images, 10 (top), 10 (bottom), 13, 19 (bottom right); J.R. Bale/Alamy, 19 (bottom left); Roger Harris/Science Source, 20

Editor: Alyssa Krekelberg
Series Designer: Colleen McLaren

Library of Congress Control Number: 2020940300
Publisher's Cataloging-in-Publication Data
Names: Bassier, Emma, author.
Title: Mercury / by Emma Bassier
Description: Minneapolis, Minnesota : POP!, 2021 | Series: Planets | Includes online resources and index
Identifiers: ISBN 9781532169106 (lib. bdg.) | ISBN 9781532169465 (ebook)
Subjects: LCSH: Mercury (Planet)--Juvenile literature. | Planets--Juvenile literature. | Solar system--Juvenile literature. | Milky Way--Juvenile literature. | Space--Juvenile literature.
Classification: DDC 523.41--dc23

Hello! My name is

Cody Koala

Pop open this book and you'll find QR codes like this one, loaded with information, so you can learn even more!

Scan this code* and others like it while you read, or visit the website below to make this book pop.
popbooksonline.com/mercury

*Scanning QR codes requires a web-enabled smart device with a QR code reader app and a camera.

Table of Contents

Chapter 1

Nearest to the Sun

There are eight planets in the **solar system**. All the planets **orbit** the Sun. Each planet is a different distance away. Mercury is the closest.

Mercury
Watch a video here!

Mercury follows an egg-shaped path around the Sun. It is the smallest planet. Mercury is a grayish-brown color.

Sometimes space experts create colored pictures of Mercury. The colors show different rocks on the planet's surface.

Chapter 2

Three Layers

Mercury has cliffs and **craters**. The craters come from chunks of space rock hitting the planet.

Learn more here!

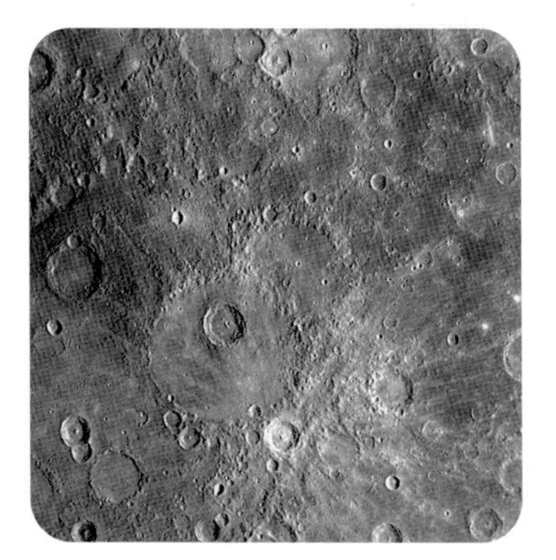

Comparing Layers

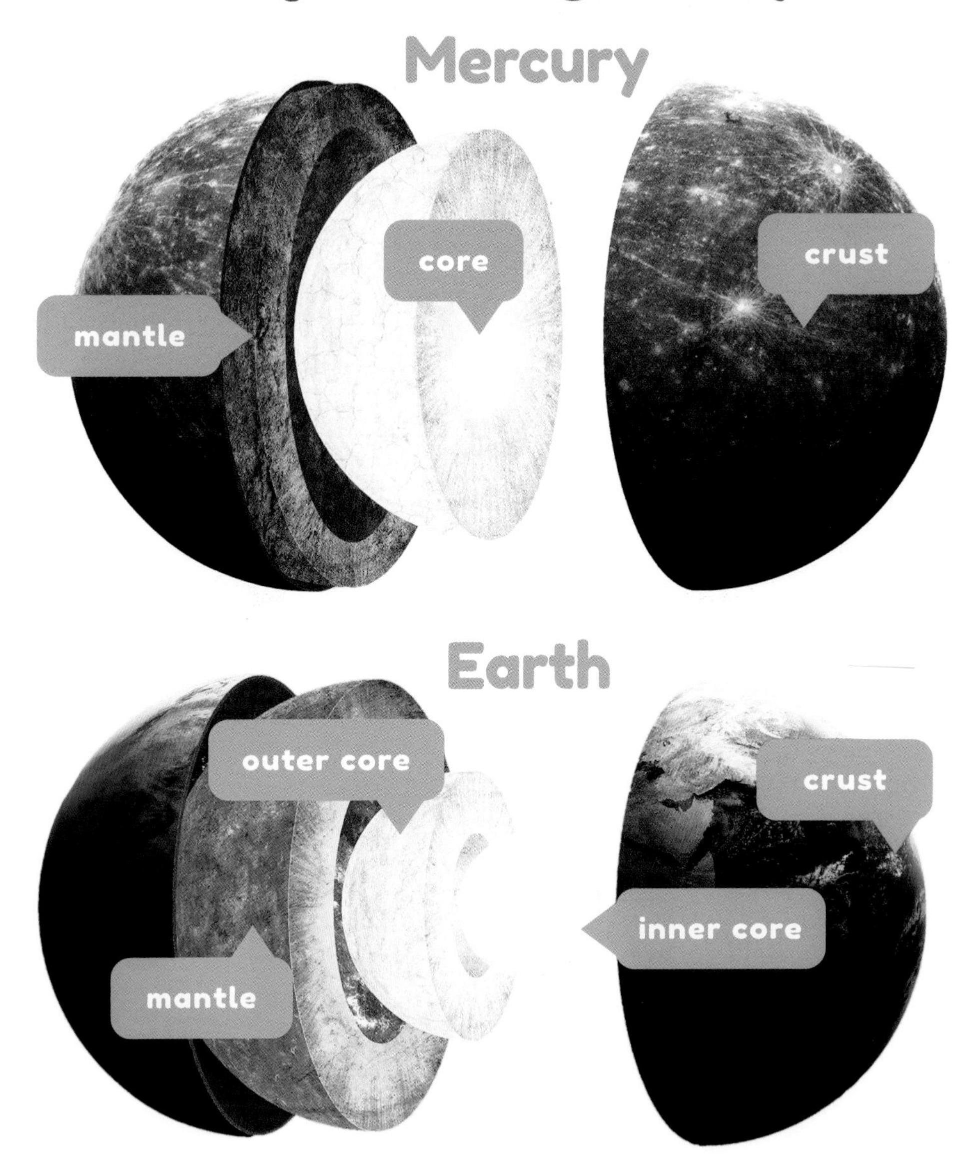

Mercury has layers. The outside layer is called the crust. It is hard. Then, there is the mantle. It is made of **molten rock**. The center of the planet has a huge solid core.

Mercury and Earth have similar layers. But Earth's core has two parts.

Chapter 3

Spinning Around

Mercury is the fastest planet in our **solar system**. It moves around the Sun in just 88 days.

Complete an activity here!

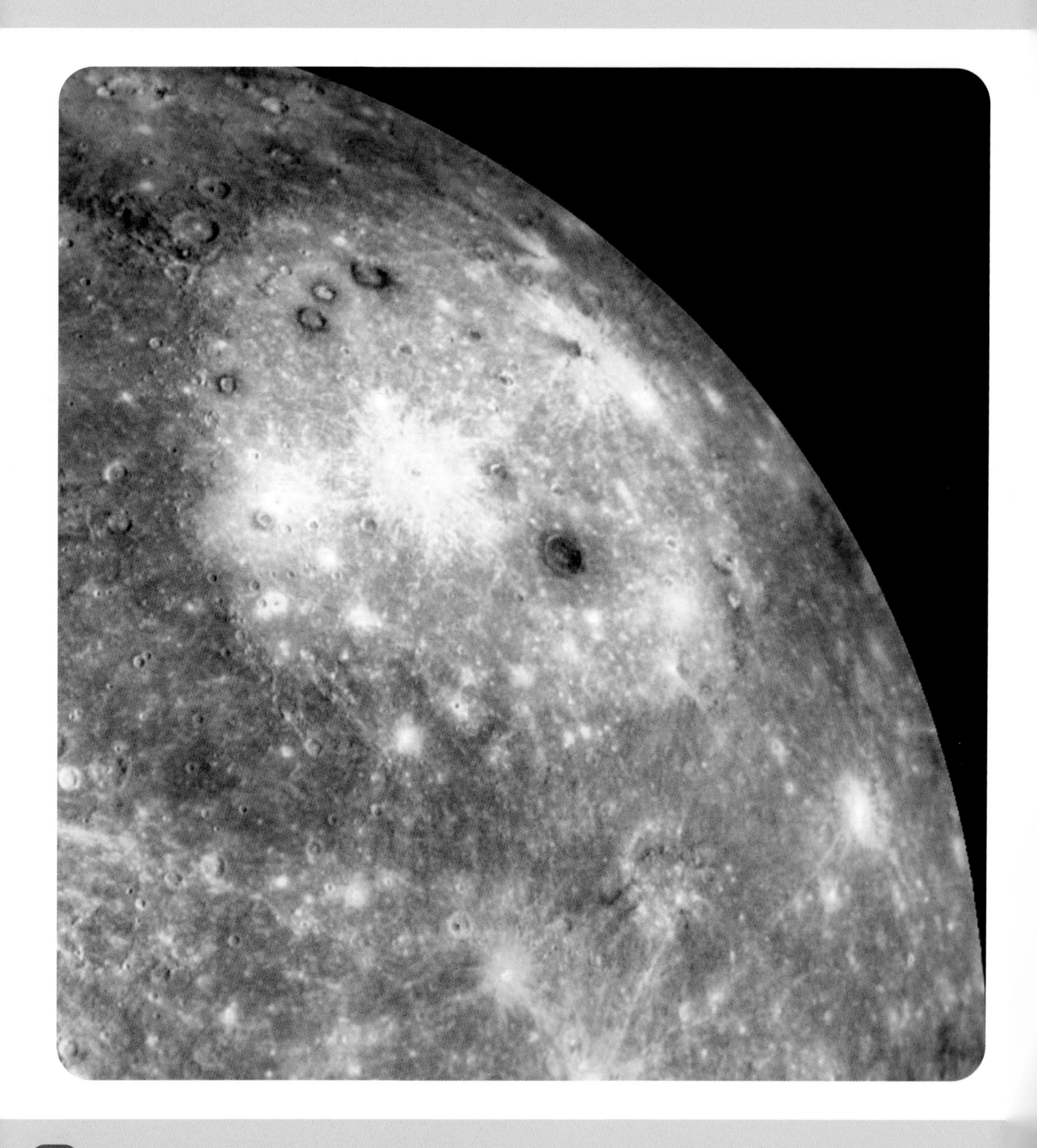

As it moves, Mercury turns on its **axis**. The planet spins slowly. It takes 59 Earth days to make a full turn. So, one day on Mercury lasts much longer than a day on Earth.

Earth's axis is tilted to one side. But Mercury's axis is nearly straight up and down.

Mercury has a strange sunrise and sunset. In some places, the Sun will rise briefly, then set, and then rise again. The opposite happens at sunset. A full day-and-night cycle takes 176 Earth days.

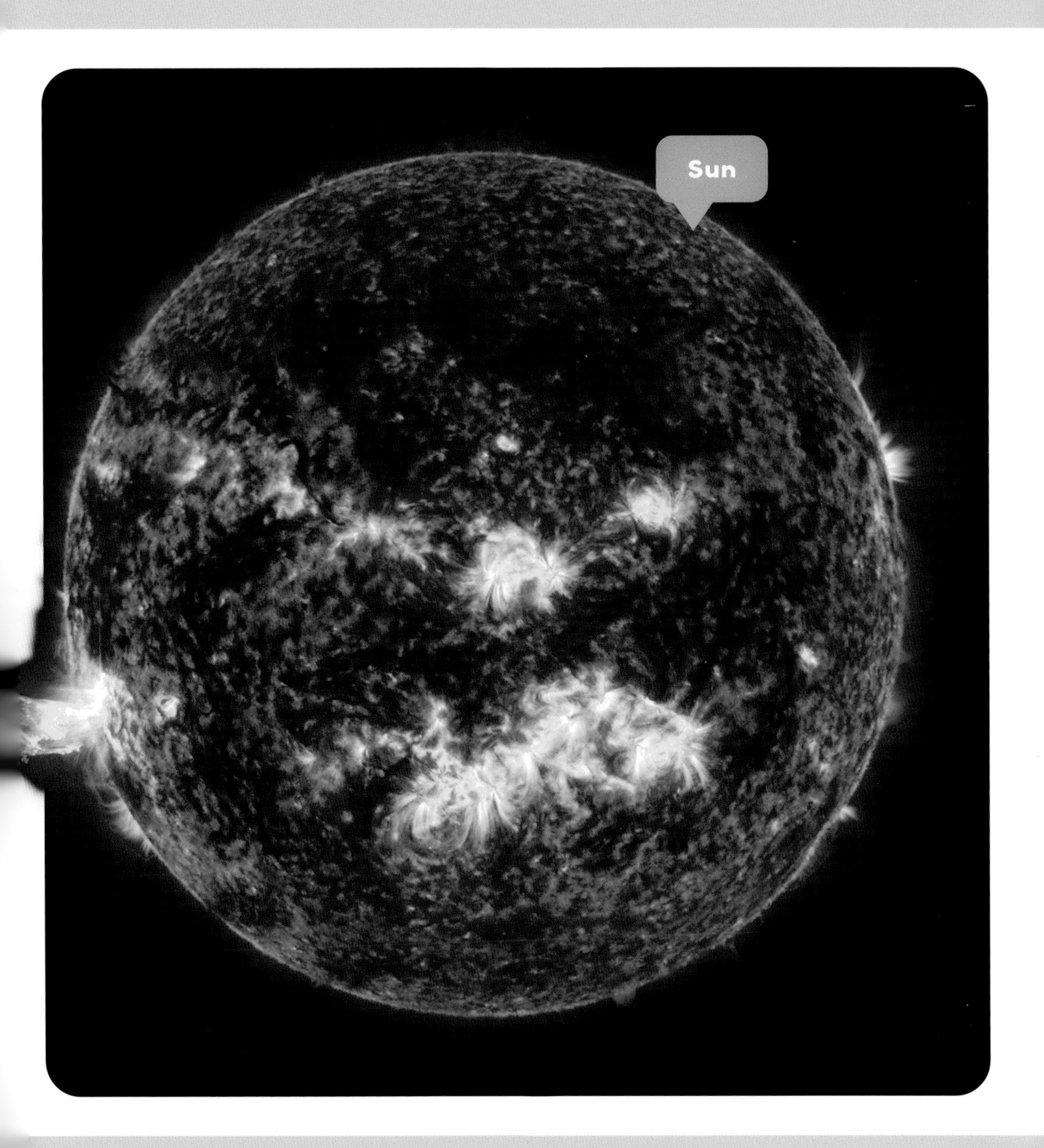
Sun

Chapter 4

Unprotected Planet

Unlike Earth, Mercury has a very thin **atmosphere**. The atmosphere does not protect the planet. Space rocks crash into it. The Sun can quickly heat it up.

Learn more here!

Because of Mercury's thin atmosphere, heat can leave the surface easily. So temperatures grow very hot or very cool. People cannot live on Mercury.

Temperatures on Mercury reach 800 degrees Fahrenheit (430°C) during the day. At night, they drop to –290 degrees Fahrenheit (–180°C).

Making Connections

Text-to-Self

Compared to Earth, Mercury has long days. If you could spend your time doing anything for a whole day on Mercury, what would you do?

Text-to-Text

Have you read other books about planets? How are those planets similar to or different from Mercury?

Text-to-World

People cannot survive on Mercury. What do people need to be able to live on a planet?

Glossary

atmosphere – the layers of gases that surround a planet.

axis – an imaginary line that runs through the middle of a planet, from top to bottom.

crater – a deep hole in the ground.

molten rock – hot, melted rock.

orbit – to follow a rounded path around another object.

solar system – a collection of planets and other space material orbiting a star.

Index

Online Resources

popbooksonline.com

Thanks for reading this Cody Koala book!

Scan this code* and others like it in this book, or visit the website below to make this book pop!

popbooksonline.com/mercury

*Scanning QR codes requires a web-enabled smart device with a QR code reader app and a camera.